iPhone 7:

How to use your latest Apple's device to the fullest (tips and tricks, hidden features)

Tailor Jacobs

Copyright © 2017 Tailor Jacobs

All rights reserved.

ISBN: 1545085013
ISBN-13: 978-1545085011

CONTENTS

Introduction iv

Chapter 1 – Design Changes 1

Chapter 2 – New General Features 6

Chapter 3 – Messages and Notifications 12

Chapter 4 – Security 17

Chapter 5 – Camera 21

Chapter 6 – Music and Sound 26

Chapter 7 – Other Helpful Features 30

Conclusion 37

Introduction

Released in September of 2016, the iPhone 7 and 7 Plus are the 10th generation of iPhone to be released by Apple, and as usual presents users with a wide array of new and upgraded features that goes far beyond the new color choices you get for the casing.

Finding all of these new features on your own can take a while, and you might end up missing out on something that could be a great use to you in your daily life. On the other hand, you probably don't want to read through the entire manual just to find the things that are new or different.

This book is designed to speed this discovery process, pointing out some of the changes so that you can check out the ones that seem interesting to you for yourself. Whether you're upgrading from a previous model or this is your very first iPhone, the tips and tricks described in this book will help you get to know your new phone.

Each chapter in this book discusses a different broad aspect of your phone, starting with general features and then delving into more specific topics like the camera or security. You can go through it front to back to get a general overview of everything your phone can do or jump to the chapter that seems the most useful.

Apple regularly releases updates for its products, so it's a good idea to check periodically and see if any new features have been released lately or are in the works. It's also a good idea to do the same thing with the app store. As third-party developers get familiar with the new model, you'll see more and more programs coming out that are designed to take full advantage of the upgrades.

Chapter 1 – Design Changes

When you first look at the iPhone 7, it probably looks like not much has changed since previous models like the 6s. Appearances can be deceiving, however; there are a lot of new features and design elements on the iPhone 7 that make it a significant upgrade, regardless of which model of iPhone you had before.

The iPhone 7 uses iOS 10 instead of iOS 9. This brings a lot of new features into the mix (explained in more detail in the next chapter). It also has an improved visual display which offers you a wider selection of colors for your screen, giving you a more accurate display of any pictures or videos you're viewing.

There are currently two versions of this latest iPhone model: the iPhone 7 and the iPhone 7 Plus. The 7 Plus is a larger and heavier phone, with a 5.5 inch display (as opposed to the 4.7 inch display of the 7). This makes the 7 Plus better for playing games or watching movies, but some users find it uncomfortably large, especially if they're accustomed to using their phone with one hand.

Perhaps the biggest difference between these two versions, however, is in the rear-facing camera. The 7 Plus has two different cameras on the back in addition to the front-facing Face Time camera. One of these has a wide angle lens while the other uses a telephoto lens. You have the option of switching between them or of using them together for an optical zoom effect; there's more info on this in chapter 5.

Water resistance

It used to be that any kind of exposure to water was deadly for your phone, leading to all kinds of glitches and failures. This could be especially frustrating since water damage is typically excluded from warranty agreements. The current generation of phones helps to limit this problem with improved water resistance.

The iPhone 7 can handle brief splashes or relatively light rainfall without taking any lasting damage. It cannot be fully submerged without damage, unlike some new phone models (such as the Samsung Galaxy S7) but it should be able to come away from a brief fall into a puddle or spray from the sprinkler unscathed.

If your phone does get water in its internal mechanisms, the phone is able to at least detect this for you and display a warning to the user. This will let you know that you should turn your phone off until it has a chance to dry so you don't cause any further damage.

Lightning port

The most controversial change from previous models of the

iPhone is the fact that the 7 and 7 Plus do not come with a headphone jack. Instead they use an all-digital connection called a Lightning Port, which you can connect to using the included AirPods, or any wireless headphones.

If you still want to use your usual wired headphones—or if you need to use the headphone jack to plug in other peripheral devices—you can still do so using the included adapter. This will usually come taped to the back of your AirPod case, so if you think you didn't receive one in your box, check there to see if you missed it.

Home button

Past versions of the iPhone had a home button that physically clicked when you pressed it down. This newest model instead uses something called "haptic feedback," that gives you the sensation of pressing the home button even though it's actually static. The feel of pressing it will be so similar to clicking that you may not even realize you're not pushing a physical button unless the phone is turned off.

One advantage of this new home button is the fact that you can change the intensity of the vibration to customize the feel of pressing the button. To make adjustments, go to the "Settings" menu, then "General" and finally "Home Button."

The fact that it is no longer a physical button does affect how you force a hard restart should your phone ever freeze. Instead of pushing the home button, press and hold the power button on the right side of your phone, and push the volume down button on the left side at the same time. Keep both held down until your phone restarts and you see the Apple logo on the screen.

Chapter 2 – New General Features

While there are some slight modifications to the exterior of the iPhone 7 as compared to previous models, the majority of the improvements were made to the features and functionality of the operating system—including the introduction of brand new ways of interacting with your phone.

One exciting feature of the iPhone 7 is that it allows you to delete the things that come installed on your phone that you don't use. In the past, most people would have a folder on their iPhone that contained all the apps that came installed on it that they don't use. As of iOS 10, however, you can delete things like the iTunes store or the stocks app. This can let you free up more space for things you actually want on your phone, like pictures, music, or videos.

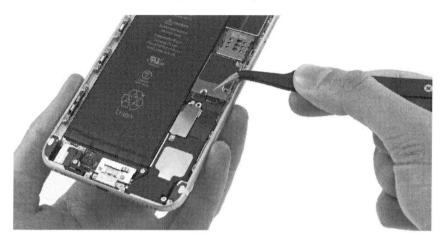

The Control Center in iOS 10 received significant updates over iOS 9. You can bring it up by dragging your finger up from the bottom of the desktop. You can jump straight to a variety of popular apps from the Control Center, but it's also the place to find some other features most users will find very important.

The Control Center is where you can toggle features like your WiFi or Bluetooth connectivity on and off. It's also where you find controls for AirDrop (a program that lets you exchange files with any other iOS device) and AirPlay (which can send your iPhone's screen to another iOS device, like Apple TV).

If you're listening to music when you open the Control Center, the media screen is the first thing that'll come up. If you're not, you can still get to your music controls from this screen; just swipe left across the Control Center to get media commands like play, pause, or stop, or to jump back to the last media app you were using.

Force touch

Perhaps the most potentially useful of the new functionality of your iPhone 7 is the addition of force touch or 3D touch. The touchscreen on these phones is sophisticated enough it can detect different levels of force and pressure that are applied to the screen, giving you different options depending on whether you press hard on the icon or simply tap.

Force touch can take some getting used to if you're not accustomed to thinking about how hard you push on the screen when you select things, but it gives you a new way of interacting with your apps without having to click through a lot of different menus. It's also app specific, meaning it'll do different things depending on which program you're using at the moment.

When you're on your lock screen, a force touch brings up the last few messages you've sent and received in a given conversation, and lets you type a new message so you can chat without unlocking your phone. A force touch on the Mail icon gives you a menu of options including inbox, search, and new message; doing it on the Photos app will bring up your photo folders.

When you're on your home screen, you can use force touch to open up your app switcher (the portal that lets you view all of your currently open apps). Press down on the left edge of the screen then move your finger toward the center. You can also switch to the most recent app you were using by force touching the left edge, then gently sliding your finger to the right.

The force touch option does some unexpected things with smaller applications on your phone, as well. A force touch on your flashlight app gives you the choice to change the brightness (low, medium, or bright are your options). Pressing on the timer brings up commonly used timers, while a force touch on the message icon brings up quick access to new messages and a shortcut to your most frequent contacts. All of the icons in the Control Center are force touch sensitive, as well, bringing up sub-menus when you press them.

Outside apps have the ability to integrate force touch technology into their projects, as well. Instagram, for example, gives you options like new post, search, or your activity page when you force press within the app. Though not many app developers have started to take advantage of this feature just yet, that's sure to change in the future.

Raise to wake

As of iOS 10, you don't need to unlock your phone to wake it up. The screen will flare back to life as soon as you pick up your phone, and though you'll still need to unlock it to use a lot of your features, you can do a lot more from the lock screen than you used to be able to, letting you do things like check your notifications, reply to messages, or open your camera app without any extra clicking.

This might sound concerning to you from a security perspective, but don't worry—you can't get into anything sensitive without unlocking your phone first. For example, you can take new photos but not look at old ones until you've unlocked your phone. If you're still concerned, you can turn this feature off. Go to "Settings" then "Display and Brightness," where you'll see a toggle that says "Raise to Wake."

Bedtime mode

Many people have been using their phone as an alarm for years, and the bedtime feature takes this to the next level. Instead of simply setting an alarm, this feature lets you determine how many hours you want to sleep. If you've set an alarm already, it will give you an alert telling you when it's time to go to bed so you can get the sleep you want. It will also track your sleep through Apple Health to make sure you're getting the right amount and quality.

You can turn on bedtime mode by going into your clock app, where you'll see this feature listed at the bottom of your display. You can then determine how much sleep you want to get and which days you want to have an alarm. You can also change the sounds it uses for alerts or how far before your designated bedtime you want it to give you a reminder.

Chapter 3 – Messages and Notifications

Smart phones are capable of doing such a wide array of different things that it can sometimes be easy to forget that their main purpose is communication. Thankfully, Apple did not forget about this key aspect of your iPhone when they were designing the 7 and 7 Plus models, and you've got some excellent new options for both your messages and your notification center.

The fact that you can see your notifications and reply to messages without having to unlock your phone is a huge time saver, letting you simply glance at your phone to check on these things rather than having to type your way through a variety of menus.

If you just get too many notifications, the improved notifications center on iOS 10 also allows you to clear all of your old notifications with the push of a single button. On previous models, you had to clear your notifications day by day. On the iPhone 7, you can simply force touch the "X" at the top of the list and it will bring up the option "Clear All Notifications."

On the other side of things, if you feel like you miss a lot of notifications, you can have your phone make them more obvious when they come in. Go to the "Settings" menu, then "General" and "Accessibility." You'll see a toggle there that says "LED Flash for Alerts." If you toggle this, the camera's flash will ping until you check your notifications. Keep in mind this will cause your battery life to drain more quickly.

One of the most-advertised changes to the messaging capabilities on iOS 10 is the addition of flash message effects, like confetti accompanying a congratulations message or balloons that help you wish someone happy birthday. If you don't want these to come up on your phone, though, you can turn them off by going to "Settings" then "General" and "Accessible," where you'll see a toggle for "Reduce Motion." If you're not getting those effects and you want to, check here to see if the feature was accidentally turned on.

Custom missed call messages

When your phone gets a phone call and you don't answer it, you can set your iPhone to deliver a text response to the person who called you with a single tap—perfect if you're in a meeting and can't take the time to type a whole message. There are a few stock responses that come included in iOS 10 but you can also create your own. Just go to "Settings" then "Phone," where you'll see the option "Respond with Text." You can type whatever you want here, then send that message with a single tap when you miss a call in the future.

Hidden messages

Sometimes you want to send someone a message that's for their eyes only. If you want to make sure no one can accidentally read a message before the owner of the phone can get to it, you can type your message and then press down on the blue up arrow. This will bring up the effects menu; select "Invisible Ink." It will look like static or sparkles on your end, but will send to your recipient as a hidden message, the text of which won't be revealed until they tap or swipe it.

Hand-written effect

If you want to draw or hand-write some aspect of your message, you can now do so on your iPhone. To find the button that will let you do this, turn your phone sideways while you're in your messages interface. This will bring up an expanded keyboard, one of the keys in which looks like a squiggly line with a loop. Tap that button to enable hand-written text or images in your message, then use a stylus (or your finger) to draw out what you want to send.

Etch-a-Sketch erasing

If you played with an Etch-a-Sketch as a kid, you know one of the coolest things about the toy was being able to shake it to make the screen erase. You can now do the same thing with your phone. If you've typed a long message and need to delete it, you don't have to mash your backspace key to death. Just shake your phone swiftly. A message will pop up asking if you want to delete the text you've typed; select yes, and it will disappear.

Chapter 4 – Security

With how much of our lives exist on our phones in the modern day, the security of your iPhone is ever more important. Apple is continuously improving the security features on their devices, and their most recent iPhone upgrade utilizes some advanced technology to make sure your phone is secure.

One of the big concerns people have with their phones is how much information is being collected about them, and how that information could be used to track their activities. If you want to check on how much your phone knows about you, go to the "Settings" menu and select "Privacy." There's a lot of information here, including "Location Services," which keeps track of your phone's location (and, by extension, yours). You can turn off tracking in this menu if you'd like.

The ability to do certain things on your phone without entering a passcode to unlock it is concerning to some users. As mentioned in chapter 2, you can turn this feature off easily by going into the "Display and Brightness" sub-menu of your "Settings."

Touch ID

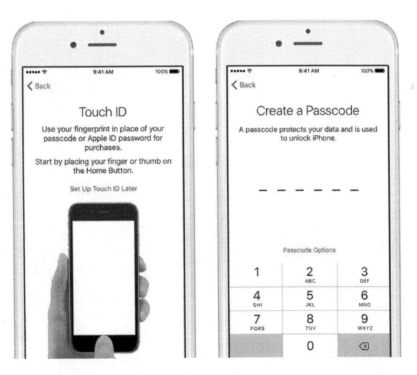

One way Apple uses advanced technology to protect your phone and data is with their new Touch ID feature, which uses fingerprint sensors to make sure that you're the only one who's able to access all the features and information you have on your device.

You can set your phone to unlock for your fingerprint in one of two ways, either with a heavy press or a light swipe. You can switch between these options by going to "Settings" and then "General" and "Accessibility," where you'll see the "Home Button" option. Toggle the switch that's labeled "Rest Finger to Open." When it's off, you'll need to press to unlock; when it's on, you can unlock with a swipe.

You can also use Touch ID to verify purchases made through your phone on select sites—including iTunes, the app store, and any purchases made with Apple Pay—preventing anyone who's borrowing or using your phone from also spending your money. To activate this, go to "Settings" then "Touch ID and Passcode." Here you can determine which purchases require a Touch ID fingerprint.

Restrictions

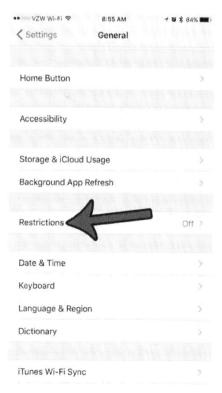

This particular feature will be of particular interests to parents who sometimes allow their children to use their phones (or whose children sometimes help themselves to use of their phones). You can get to this in the same place you find the Touch ID settings, and there are a wide variety of things that can be restricted.

When you establish a restriction,

you set up a PIN-protected menu that lets you disable features like AirDrop or Siri. You can also require a password for installing or deleting apps, restrict which websites the phone is able to browse, or even prevent the download of viewing of movies above a certain rating.

Chapter 5 – Camera

The camera on the iPhone 7 has been markedly improved over the camera that came on past models of the phone. It gives you a 12 megapixel sensor that includes optical image stabilization. Compared to the 6s, you get 60% faster performance and a wider color capture, as well as a lens that can bring in 50% more light, bringing extra clarity even to poorly-lit scenes.

Apple has also made improvements to the front-facing camera and the flash mechanism. The flash now puts out 50% more light and can reach 50% further, and also is able to compensate for flickering or unsteady light. The front-facing camera, meanwhile, has been increased from 5 megapixels to 7 megapixels, giving you better resolution on selfies or in FaceTime chats.

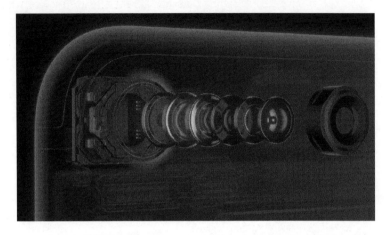

If you have an iPhone 7 Plus, your phone has the most impressive phone camera system yet, with two different cameras on the back of your phone. This enables the new Portrait mode, which gives you photos with two different layers: the person in the foreground in sharp focus and the background, which is artfully blurred. The added lens also means you can achieve a pure optical zoom of up to 2x and a partially-optical zoom up to 10x.

Of course, as with all new developments, this second lens can also cause problems, especially when you're trying to record video. If your phone tries to switch between the lenses while recording a video you may get flickering in the footage. To prevent this, you can lock the second lens in place. Just go to "Settings" then "Photos & Camera" and "Record Video," where you'll be able to toggle the "Lock Camera Lens" switch.

Instagram has already been updated to take advantage of these recent improvements to the iPhone camera technology. They've added an array of new filters that are designed to work well with the improved cameras, and the brighter color capture has been brought to the forefront. The app has also added notifications that let you see updates from your contacts without exiting whatever app you're currently using.

The force touch functionality that's been added to the iPhone also has a role in your camera app. You can use it to review the photos you've taken recently. Press down on the thumbnail of your most recent photograph (it will be in the lower left-hand corner of the app). This will open up your photo gallery. You can also lightly tap the picture and then drag your finger left and right once it pops up to brows your photos. To go back to the camera app, simply tap it again.

Live photos

The live photos option was added in the iPhone 6s, and has been improved with the iPhone 7. The biggest improvement was the addition of stabilization, which lets you get clearer images and better resolution on the video that comes before and after your still image.

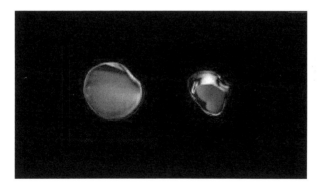

Apple has also added the ability to edit your live photos, with new tools for you to use. These same tools were also made available to third party developers with the release of the 7. Though there haven't been many people taking advantage of this yet, expect to see more editing and filter options coming out through third-party apps in the near future.

Searching with Siri

You can use geotracking (the feature that keeps track of what specific location your photos were taken in) in combination with Siri voice commands to give you a lot of different voice-activated search options for your albums. You can ask Siri to search for photos from a specific location, or can ask her to look up pictures from a given date or time frame, and can use these same search functions for videos.

Siri is also able to identify specific objects within your photos. Its ability to do this is impressive, and you may be surprised exactly what kinds of objects Siri is able to pick out of your pictures. You can ask Siri "Show me pictures of cats" and it will bring up everything in your photo album that fits the description. You can have a lot of fun playing with this feature and seeing what all Siri is able to recognize.

Video resolution

The iPhone 7 is the first model capable of shooting video in an ultra high definition format. To change the resolution of the video you shoot—or to check what resolution it's currently set to—go to the "Settings" menu, then select "Photos & Camera." Scroll down until you find the "Record Video" option.

Opening this menu gives you a variety of resolutions. The lowest is 720p, which records 30 frames per second; the highest 4,000p for an ultra-clear, super smooth recording. Be forewarned that the higher the resolution, the more memory the file will take up. A single minute of video at 4,000p takes up 350MB, about six times the space of a minute-long video at 720p.

Chapter 6 – Music and Sound

Though the decision to drop the headphone jack from the iPhone 7 was initially controversial, the end result is something most users can get behind: an improvement in the overall sound quality. The sound signal had to be converted before it could be translated through a jack. With the new lightning port, the signal is all digital, meaning you can listen to hi-res audio with no loss of detail.

The phone's speakers have also been improved over previous models, for those times you want to listen to your music out loud. They have an increased dynamic range and can produce about twice the volume of the speakers on the 6s.

The music app has been re-designed for iOS 10. It was altered to integrate the new force touch technology, for one thing. Perhaps the coolest way force touch has been introduced is in setting up a queue of songs you want to play next. Force press on any song from your library while you're listening to another track. A menu will pop up; tap the option "Play Next" to slot it into the queue. This works with playlists or albums as well as songs.

You can do more than just listen to the songs on your playlists. One example of the music app's new extra features is that it can show you the lyrics to the song you're listening to. Tap the "Now Playing" section at the bottom of the app to bring up the full song, then tap the three dots at the bottom right-hand corner. This will bring up a menu where one of the options is "Lyrics." This won't be available for all songs, but more are being added all the time.

Don't only think about the music app itself when you're looking for new things you can do with your songs. The clock app, for example, lets you set how long you want to play your music. Go to the "Timer" options then go down to the "When Time Ends" tag and switch out the command "Stop Playing" for the usual alarm. This can be especially helpful if you like falling asleep listening to music but don't want it playing all night long.

AirPods

AirPods are the newly designed wireless earbuds that come with your iPhone and are designed to work with the lightning port to connect without a plug. They're smart little devices that not only connect to your phone automatically but also can sense when you're wearing them. If you take them off, the music pauses automatically and the AirPods go into sleep mode, which conserves the battery life.

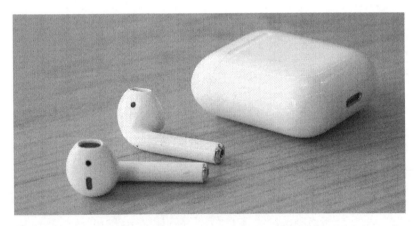

The AirPods recharge when you put them back in their case, which can hold up to 24 hours of battery life. Keeping them in their case is a good idea anyway; one of the main complaints when the AirPods were released was that they're very small and easy to lose, and a new set costs around $150.

In addition to the speakers, the AirPods contain a microphone meaning you can use them to give commands to Siri or talk on the phone. They use a unique beam technology that blocks out background noise and lets them focus on the sound of your voice—pretty impressive functionality from such a small package.

Chapter 7 – Other Helpful Features

There are loads of features and special tricks that you can use on your iPhone that you might not be aware of. Some of these are settings that make your phone easier to use, while others are tools you might not have realized were available to you.

Keep in mind that turning on or altering some settings will cause your iPhone to use more power, which can translate to a shorter battery life. You can compensate for this by conserving power in other places. Go to the "Settings" menu, then click "Battery." From here, tapping one of the apps on the list will show you how long it's been running in the background. Shutting down programs you're not using at the moment will save more power for the things you do want to use.

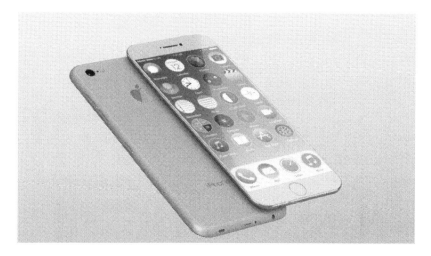

Your iPhone can also be customized in a variety of ways to suit your needs and style. This goes beyond changing your background image or screen saver. There are a variety of different widgets you can add to your lock screen, letting you look at them without unlocking your phone. This includes news, weather, and traffic info along with your contacts and conversations. The list of things you can access from your lock screen is constantly being updated.

Some of the tips and tricks that follow in this chapter are new to the iPhone 7 and 7 Plus, while others were carried over from previous models, though you may or may not have realized they were there, and may have been improved since the version that existed on your phone.

Apple Pay

This feature was available on the iPhone 6, but if you're upgrading from anything that came before that it'll be new to you. Apple Pay is the name of Apple's mobile payment system. You have to set up the wallet app to use it, but once you've entered in your payment information you'll be able to pay for things by just tapping your phone.

Not every store supports Apple Pay; if they do, you'll usually see an Apple Pay logo on display near the register. It's also useful when you're shopping online, letting you quickly pay for items purchased through your web browser. As was mentioned in chapter 4, you can set up a Touch ID lock on this feature if you're concerned about other people buying things when they use your phone.

Landscape mode

You've probably noticed that your phone can tell which way you're holding it and adjusts the view accordingly. What you might not realize is that this does more than just flip text and images so you can still read them. There are some buttons and view options you can only get when you flip into landscape mode.

Just like the force touch, landscape mode is app specific, meaning exactly what it does that's different from the standard portrait view will vary depending on what program you're using. The most common use of landscape mode is with your keyboard, which expands to give you keys for things like undo, cut, copy, or paste.

There are other programs that will give you extra buttons, as well. The hand-drawing option in the messages window that was mentioned in chapter 3, for example, is one of these, or the calculator tool, which expands into a scientific calculator.

In some apps, turning to landscape mode enables you to see more detail. The calendar is one of these, showing more information about items on your calendar. The stocks app and health app both give you more view options in landscape, expanding so you can see hourly, daily, weekly, and monthly breakdowns at a glance.

Other apps may give you a split view. In the mail and messenger apps, landscape gives you two columns, one on the left for viewing all messages which, when tapped, open on the right. Experiment with turning your phone sideways in your other apps, as well, to see if extra buttons or features are hiding in the extra space.

One-handed operation

The larger screens of more recent iPhone models are great for when you want to watch your favorite shows on the bus to work, but can be much less convenient if you need to open up an app or check your messages with one hand. If you double-tap the Home button, the top half of the screen will slide down toward the bottom, letting you reach all the icons with your thumb.

If you do this often, you'll probably find it helpful to rearrange the folders and apps on your home screen with your thumb's reach and movements in mind. Put the ones you use the most at the bottom right (or bottom left, if you're left-handed) where your thumb will have the easiest access.

Screen filters

If you're color blind, the iPhone offers four different filter variations to accommodate different manifestations of the ailment: grayscale, red/green, green/red, and blue/yellow. You can get to these options by going to "Settings" then "General" and "Accessibility," followed by "Display Accommodations," where you'll find the option "Color Fiters."

This menu also lets you put a tint of a specific color over your entire screen. You can customize the specific hue and intensity. Even people who aren't color blind may find a use for this; Night Shift mode puts a blue filter over your screen so it's easier on your eyes in the dark.

Handy tools

Did you know your iPhone can function as a level? When you swipe your finger left over the screen in Apple's compass app, it brings up a virtual level so you can easily see whether you're hanging a new picture straight or not.

The magnifier tool is also extremely helpful around the house. It basically takes advantage of the zoom on your lens without taking a picture, letting you see things in more detail on your screen. Use the white circle button to capture the image. Once you do, you'll be able to zoom and navigate to get a better look at what you're trying to see.

You can activate this feature by going to "Settings" then "General" and "Accessibility," where you'll find the toggle option "Magnifier." After you've turned it on, it'll come up when you triple-click the home button. If this doesn't work, go back to the "Accessibility" menu, and this time go to "Accessibility Shortcut," where you can set that command to the magnifier.

Conclusion

Every time a new version of the iPhone is released it takes a little bit of getting used to, especially when the upgrades change the way you interact with your screen and home button. The iPhone 7 got only a luke-warm reception when it was first released, but as people have had more time to get accustomed to its new feel, a lot of users are starting to come around.

The iPhone 7 and 7 Plus work great in conjunction with other Apple products, like the Apple Watch or Apple TV. You can send a show from your phone to your TV screen when you get home, or use the Find My iPhone function on your phone to look for your watch (or vice versa).

Though it looks relatively unchanged, the upgrades that Apple put into place on the iPhone 7 impact every aspect of your phone's operation. The improved cameras—and addition of the second camera on the 7 Plus—give you even more options for your pictures than you had before, while the improved screen resolution lets you view them more accurately.

The changes to the physical construction are also vast improvements. The increased water resistance helps prevent irreparable damage to your phone. Not only does the switch from the headphone jack to the lightning port improve sound quality in your headphones, but your AirPods can be used for hands-free communication with Siri.

Take some time to play with what kind of information you want to be able to access from your lock screen. Having things there like your messages or info on the traffic and weather lets you check on them at a glance, without having to unlock your phone. While additions like message expressions may be more fun to test out initially, you'll likely find the changes to the lock page and home screen to be the most impactful on your day to day life.

The iPhone is a deceptively powerful machine, giving you access to a variety of information and tools right in the palm of your hand. I hope the information included in this book has helped you to better navigate your new phone and take full advantage of its features, both old and new.

Thank you for reading. I hope you enjoy it. I ask you to leave your honest feedback.

I think next books will also be interesting for you:

How to Install Kodi on Firestick

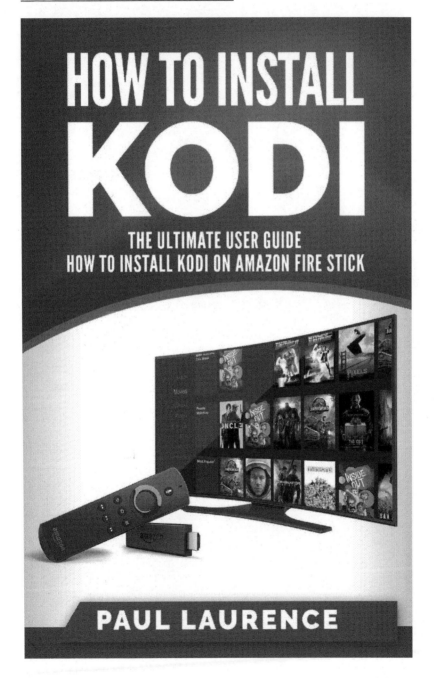

Amazon Echo

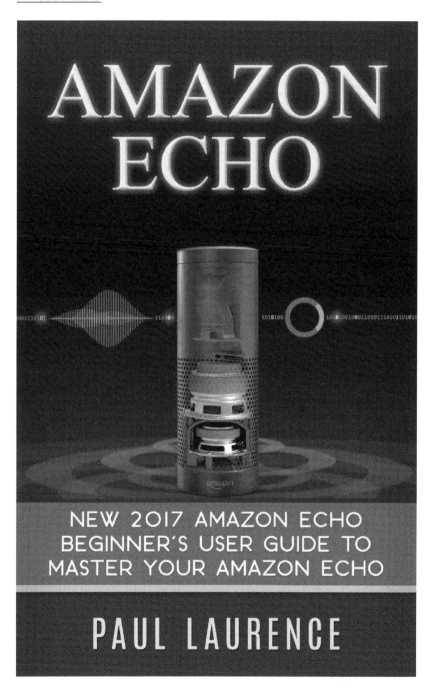

Amazon Fire TV

Made in the USA
Middletown, DE
15 April 2019